WEIRD ANIMAL SCIENCE

ANIMAL CAMOUFLAGE

THE BREAKDOWN

VIRGINIA LOH-HAGAN

45TH PARALLEL PRESS

Published in the United States of America by Cherry Lake Publishing Group
Ann Arbor, Michigan
www.cherrylakepublishing.com

Photo Credits: © SofieLion/Shutterstock, cover, 1; © bsd studio/Shutterstock, cover, 1, 13; © Margaret M Stewart/Shutterstock, 6; © JosepPerianes/Shutterstock, 8; © Anna-Carina Nagel/Shutterstock, 9; © KenCanning/iStock, 11; © lexey Seafarer/Shutterstock, 12; © Julia Remezova/Shutterstock, 14; © Noorhussain/iStock, 16; © GranTotufo/Shutterstock, 19; © Scheferson/iStock, 20; © Polina Tomtosova/Shutterstock, 21

Graphic Element Credits: Graphic Element Credits: Cover, multiple interior pages: © WindAwake/Shutterstock; © Theo_hrm/Shutterstock; © Rodin Anton/Shutterstock; © Petr Vaclavek/Shutterstock; © Nikolaeva/Shutterstock

45th Parallel Press is an imprint of Cherry Lake Publishing Group.

Library of Congress Cataloging-in-Publication Data has been filed and is available at catalog.loc.gov

Cherry Lake Publishing Group would like to acknowledge the work of the Partnership for 21st Century Learning, a Network of Battelle for Kids. Please visite Battelle for Kids online for more information.

Printed in the United States of America

Note from publisher: Websites change regularly, and their future contents are outside of our control. Supervise children when conducting any recommended online searches for extended learning opportunities.

Dr. Virginia Loh-Hagan is an author, university professor, and former classroom teacher. She's currently the Director of the Asian Pacific Islander Desi American Resource Center at San Diego State University. She loves science-fiction stories. She lives in San Diego with her very tall husband and very naughty dogs.

TABLE OF CONTENTS

BREAK IT DOWN

1. DECODE IT

- Look at each letter.
- Look at letter patterns.
- Blend sounds.
- Try other possible sounds.

2. BUILD VOCABULARY

- Listen to the word.
- Match the sounds to a word you know.
- Look for context clues for words you don't know.
- Look up the meaning in a glossary or dictionary.

3. MAKE CONNECTIONS

- Name other examples.
- Draw the idea.
- Compare ideas.
- Ask questions.

TEST IT OUT

WORDS YOU'LL LEARN

Look over the list of vocabulary words. Are there any you already know?

camouflage
conceal
disruptive
mimicry
pigments

GET STARTED

When consonants are next to each other in long words, it can be a clue about how to read the word.

- Look at the letters.
- Chunk long words to read them.
- Keep consonant blends together.
- Divide between consonants that don't blend together.

The word *conceal* has 2 consonants in the middle. It is pronounced kuhn-SEEL.

- The 1st vowel says "uh": *kuhn-*.
- The 2nd vowel sound is a long *e*: *-seel*.

Sometimes animals hide. They don't want to be seen.

They **camouflage** to protect themselves.

Camouflage means blend into the environment.

It helps them trick predators.

DECODE IT

The word *camouflage* has a consonant blend near the end. It is *fl*. Keep the blend together.

- Break it into chunks: cam-ou-flage.
- The first vowel sound is a short *a*.
- The next vowel sounds says "uh."
- The last vowel sound says "ah."
- *Camouflage* is pronounced KAM-uh-flahj.

It helps them hunt. They mislead prey.

They attack them by surprise.

Animals have different ways to camouflage.

Some animals **conceal** themselves.

Conceal means to hide or blend in.

BUILD VOCABULARY

Conceal (kuhn-SEEL) means to hide or blend in.

- Animals do this to camouflage.
- This helps animals stay safe!
- Other forms: concealable, concealing, concealment

Pigments are substances that color skin or fur.

These pigments react to changes in temperature.

Some animals have special markings.

These markings trick the eyes.

They're **disruptive**. Disruptive means to get in the way of.

DECODE IT

The word *disruptive* has no consonant blends. Divide it between consonant pairs.

- Break it into chunks: dis-rup-tive.
- The first and last vowel sounds are short *i*.
- The middle vowel is a short *u*.
- *Disruptive* is pronounced dis-RUHP-tiv.

Zebras have stripes on their bodies. These stripes distort their outlines.

It confuses predators.

MAKE CONNECTIONS

Think about animal markings.

- What other animals have stripes or other patterns on their fur?
- Are those animals predators or prey? How do you think their stripes or patterns help them?

It makes it hard to pick out just one zebra.

Some animals copy nature. This is called **mimicry**.

An example is walking sticks. Walking sticks are bugs. They look like sticks.

Predators think they are sticks and leave them alone.

PUT IT TOGETHER

Detail:
Animals use camouflage to catch prey.

Detail:
Animals use camouflage to hide from predators.

Detail:
Some animals have special markings to camouflage.

Main Idea:
Animals use camouflage to hide and hunt.

Detail:
Some animals use mimicry to camouflage.

Detail:
Some animals change colors to camouflage.

LEARN MORE

BOOKS

Jenkins, Steve. *The Animals Book*. Boston, MA: HMH Books for Young Readers, 2013.

Lerwill, Ben, and Sarah Walsh (illus.). *WildLives: 50 Extraordinary Animals That Made History*. New York, NY: Atheneum Books for Young Readers, 2020.

Watson, Galadriel, and Samantha Dixon (illus.) *Running Wild: Awesome Animals in Motion*. Toronto, ON: Annick Press, 2020.

ONLINE

Search these online sources with an adult:
- National Geographic Kids | Animals
- Britannica Kids | Animal Kingdom

GLOSSARY

camouflage (KAM-uh-flahj) to blend into the environment

conceal (kuhn-SEEL) to hide

disruptive (dis-RUHP-tiv) to get in the way of

mimicry (MIH-mih-kree) the act of copying

pigments (PIG-muhnts) substances that add color to something

INDEX